Extra-Universal Mechanics
And the Nature of Reality

By Curtis R. Crim, BA, CEO

ISBN: 978-0-9833732-6-1

Printed in the United States of America

First Printing

Dedication

This book is lovingly dedicated to Spinky, Pepper, and Squeaky, without whom my journey in space and time would be meaningless.

TABLE OF CONTENTS

Introduction:
The Subjectivity and
Objectivity of Consciousness

This book addresses issues of physics, and those that cannot be separated from physics.

Although the discipline of physics is by definition a scientific examination of the universe, it also attempts to explain the issue of how the universe came into existence, and what might be outside of the universe.

Physics has not yet adequately explained exactly how and why the universe came into being and it utterly inadequate for addressing the issue of the nature of that which is outside the universe - the "extraverse", if you will.

Science uses inductive reasoning and an empirical approach to gathering data, and attempts to objectively come to conclusions based on actual proof and evidence that reality bears out the conclusions and theories that are the result.

However, the concept of objectivity itself is subjective, merely because it is a concept - a construct of consciousness. Further, any and all observations made by a conscious being are also subjective in nature. Any consciousness existing in our apparently three dimensional universe by its very nature can only see and cognize reality in a subjective manner.

The only way to be truly objective is to examine the universe from the *outside* of the universe, which ironically is also the *most* subjective perspective of all.

Further, science, mathematics, and physics are all culturally created concepts, therefore are also entirely subjective. Science is not an objective discipline, but rather an *attempt* to approach understanding through objectivity.

In my opinion, one of the major drawbacks of science in general is that it attempts to remove the subjective from the method, where it is impossible to do so.

I propose that a discipline needs to be developed which is based upon the science of physics, yet attempts to reintegrate consciousness itself into the equation.

The science of physics examines the universe to gain an understanding of the universe and reality. The disciplines of mathematics and physics have been created by subjective beings inside the universe. The data and observations are made and acquired inside the universe. This is a valid approach and has served physicists well, mostly, up to this point, but it also breaks down when attempting to explain the source of the universe, and what is outside of the universe.

In this book, I am taking the approach of starting out by examining and describing what is outside the universe first. I believe that by learning about various features of reality that are not in our universe, we can learn more about our universe, and gain a better understanding of our perspective of the universe. Some of the features of reality which I will discuss are other universes, membranes, the multiverse (and multiverses in general), the omniverse, the extraverse, time, and consciousness.

In order to gain any understanding of reality, it is important to understand the nature of time, not just in the universe, but also outside of the universe. After all, the source of space, which we observe inside the universe, must also be the source of time.

Chapter 1: Time

It is a commonly held belief that time is a function of, and no more than a function of, what is apparently three dimensional space.

I would like to present a different view of the nature of time.

First, consider the following analogy using the electromagnetic field:

Our ocular organs are capable of perceiving only a small bandwidth of the electromagnetic field. In other words, what we see as light is only a small bandwidth of a much larger analog spectrum.

This is similar to our ability to perceive time.

One of the problems with being a human is that our eyes present a reality to our consciousness which is highly compelling, and it is our inclination and indeed our very nature to believe that what we see around us is real.

We have evolved ocular organs which see out into what is an apparently three dimensional space, and we perceive the passage of time based on our apparent movement through said three dimensional space. However, the time that we perceive is also that which we have evolved to be able to understand.

Time also exists *outside* of the universe, but its nature is very different than what we perceive as time inside the universe.

Outside the universe, time is actually a large apparently three dimensional spectrum, of which only a very small bandwidth is perceivable to us as human beings.

This fact is due to the nature of the apparently three dimensional space that we seem to be living in within the universe.

Space, by its nature, works as a filter to allow into the universe only the part of the spectrum of time with which it is compatible.

When the nature of space is different, so is the nature of time.

We know from quantum physics that space is not simply three dimensional. In fact, our universe appears to contain three large dimensions of space, and eight small "curled up" dimensions of space.

Due to the nature of small curled up dimensions of space, we humans are not able to perceive them visually.

Since all kinds of space have time associated with them, we must also consider that the eight small curled up dimensions of space also have associated with them small curled up dimensions of time, which we are equally unable to perceive.

However, we should not conclude that these small curled up dimensions of space do not play a role in other universes.

A small curled up dimension of space which cannot be perceived in our universe might very well appear to be a large dimension of space in another universe in our multiverse.

If this were the case, then that dimension of space would also have a dimension of time which would be perceivable to beings living in that universe.

In fact, the mere existence of small curled up dimensions of space (and their associated dimensions of time) strongly suggests the existence of other universes.

Chapter 2: Other Universes

A common theme in many science fiction stories is that there are other universes, ones in which we might see a universe very much like this one, but which vary from the reality which we experience. For instance, Spock has a beard and Captain Kirk is evil.

Science fiction writers have speculated that another universe could exist that resembles this one in nearly every way, and that there are versions of each of us in said universe, but perhaps the color of your hair or eyes is different in that universe.

I believe that there is some confusion here between other universes, and other quantum realities in this universe.

These versions of other universes assume that all universes have three large dimensions of space, and time that appears to be linear as it is in this universe. I believe that this is a mistake as well.

In every universe, every possibility of what can exist *does* exist. This is where quantum realities come into play. In fact, one's consciousness is capable of shifting between quantum realities in this universe.

Sometimes, the difference between two quantum realities is so small, that it is imperceptible to a being shifting between them. Other times, the difference is major, which sometimes makes things which we know to be impossible appear to actually happen.

I would like to use my own life as an example of a quantum reality shift. I live in a reality where it is not possible for a woman to agree to go out on a date with me. It is *literally* impossible, much to my chagrin, for a woman to say the word "yes" to me if I ask her out on a date. However, if a woman was to actually go out on a date with me, then it would constitute a shift into a different quantum reality. I am not exactly sure when the shift actually would have happened, but I am certain that if I were ever to be sitting at a restaurant with a woman who is my date sitting across the table from me, that a quantum reality shift would have occurred, and I would no longer be in the same quantum reality in which I currently reside. One might speculate at this point that the quantum reality in which I live is not a very pleasant place to be.

When people talk about other universes, they are generally talking about different quantum realities in *this* universe.

The thing about other universes is that they are actually different from our universe. In fact, other universes are very, *very* different from this one.

In the omniverse, it is likely that every possibility actually occurs.

Universes come into existence which have a completely and totally different nature than that of our universe.

When a universe comes into existence, space manifests itself in one of many possible configurations. Some universes have more large dimensions of space, and fewer small dimensions of space.

It is even possible to have a universe manifest in which dozens or hundreds of large and small dimensions of space exist. The nature of such a universe, and the beings which would evolve in it, would be quite different from those in the universe with which we are (partially) familiar.

For the sake of the argument of this discussion, I am going to assume that the total number of spacial dimensions is a constant in our multiverse. In other words, another universe in our multiverse might have five large dimensions of space, but it would then have only six small curled up

dimensions of space, so that the total
number of spacial dimensions would still
come to eleven, just as it does in our
universe.

Let's examine what this other universe with
five large dimensions and six small curled
up dimensions of space might be like.

The nature of space is to filter the spectrum
of time in such a way that only compatible
time can exist inside the universe. Only the
large dimensions of space play a role in
filtering the spectrum of time in this way.
Small curled up dimensions of space have
no effect on the time that exists in a universe.

A universe with five large dimensions of
space would filter the time spectrum in such
a way that only time that is *compatible* with
five dimensional space could exist in it.
Beings who evolve in such a universe might
very well develop ocular organs (or their
equivalent of ocular organs) which would
allow them to perceive or somehow "see" in
five rather than three dimensions of space.
Further, they would also have to develop
consciousness that is capable of
understanding a reality which exists in five
large dimensions.

The nature of time would be very different than what we perceive as time in our universe of three large dimensions.

In our universe, time generally appears to be linear. In other words, it appears to be a line, thus most people are familiar with the term "timeline". In our universe, time also sometimes appears to be a point. For instance, when mass is accelerated to infinity, such as in a black hole, time appears to be reduced to zero.

If one can imagine a scenario where time is only a point and reduced to zero, it is obvious that there can be no movement in space. Nothing can move in an environment where mass is infinite and time is zero. Thus, I conclude that a black hole is *not* a wormhole; therefore transportation through a black hole is not possible.

I am digressing, but I also want to point out that if one accelerates time to infinity, space cannot exist, and thus the universe would come to an end. By definition, the acceleration of time to infinity would bring about the end of time; therefore also bring the universe to its end.

The nature of time is dependent upon the nature of the space in which it resides. The more dimensions of space which exist in a universe, the more dimensions of time also exist. We have three large dimensions of space, and time appears to have only one dimension (or none at all, at infinite mass).

However, if a universe had, for example, four large and seven small curled up dimensions of space, then time would appear two dimensional rather than one dimensional. The time in this hypothetical universe would come from a different part of the spectrum of time than time which exists in our three dimensional universe.

Beings in such a universe would evolve to be able to understand two dimensional time, and perceive time as a plane rather than a line. If they develop a lingual means of communication, they might have a word that would translate as "timeplane", which would be their equivalent of our word "timeline".

In a universe where time appears to be a plane, it might be far easier for beings to shift between quantum realities. They might indeed evolve to be able to perceive and control their movement through quantum realities as easily as we move through space. One could even speculate that they might develop a technology that would allow a service for which a company would charge a person for transportation to a specific quantum reality to which they would like to go, similarly to the way we might pay a service to bring us to a different location in space via ground or air transportation technology.

Another quality of time is that it flows at a different rate based on the number of its dimensions. I speculate that time might flow more slowly if it is a plane rather than a line. In a universe with more dimensions of space, and therefore more dimensions of time, time would flow more slowly than it does in our universe, if it were possible to examine and compare the nature of these different kinds of time simultaneously. Of course, the beings living in this other universe would not be aware of the difference; to them it would just be the time which they normally perceive.

If time were to flow more slowly in a different universe, then it might also be that it would take more energy to move through space in such a universe.

If one were a being living in a universe with four large dimensions of space, it might require a great deal more energy to move through space than it does for us in our universe. If this were the case, then it is likely that evolution would also be a much slower process, from our perspective.

Perhaps we should consider ourselves lucky that we live in a universe that has only three large dimensions of space! Otherwise, we might not even exist yet! I find some humor in this notion. On the other hand, in some other quantum realities in this universe, we probably do not exist yet, and in others our existence has already come to an end. One can even shift into a reality in which one does not exist. It happens all of the time. We call it "death".

Now that I have discussed some of the possibilities as to what other universes might be like, you might also be wondering where universes come from, why and how they come into being, and why they might share common properties, even though they are very different in nature. Therefore I will segue into a discussion about our multiverse, which is our tiny little corner of the omniverse.

Chapter 3:
Membranes and Multiverses

Our multiverse is comprised of more than universes. It is thought that our universe resides upon the surface of what is called a "membrane", or "brane" for short.

The documentaries that I have seen about the universe visually depict the membranes as planes which exist in a parallel orientation to one another. The current thinking of modern day astrophysicists is that universes manifest on the surface of these membranes. They suggest that membranes are not static, but that they move around, and that when a collision occurs, universes are created (and probably sometimes destroyed.)

I speculate that the membranes themselves are fields of pure energy, and when they collide, large amounts of energy is released, which causes the creation of universes.

Although the membranes are shown as two-dimensional planes which are parallel to each other, I want to point out that the concept of the plane is part of our discipline of geometry, which we have created and envision in three dimensional space. What we see as geometry might very well not apply to anything in the extraverse.

Because we are conscious beings who see around us a universe that has three large dimensions, we tend to think in terms of three dimensional space and three dimensional geometry. Our consciousness has evolved to understand the environment around us in this way, and it is hard for us to envision situations or environments that are very different in nature.

Depicting membranes as planes works well as a visual aid to help people understand them, but any reference to a three-dimensional construct might very well not apply to anything outside our apparently three (large) dimensional universe.

That being said, I would like to suggest that membranes might very well be concentric and sometimes overlapping spheres, rather than planes. Or that the spheres are so big that they only appear to be planes if you see them from up close.

In fact, the membranes might not be any regular geometric shape of any kind. Membranes might actually be shaped highly irregularly.

I do agree that it is the collision of these membranes that causes the creation and sometimes destruction of universes.
If the collision of these membranes does create universes, it is likely that they contain large amounts of some kind of energy.

I would also like to suggest that membranes can not only collide, but can remain in contact with each other, overlapping and creating a new set of properties which will effect the nature of the universes created in that part of the multiverse.

Our multiverse is comprised of two or probably more membranes, upon which our universe is one among probably countless others. If the constant total number of spacial dimensions for any universe is eleven (three large and eight small, in our case), it is likely that this is not a constant in other multiverses.

In other parts of the omniverse, there are different numbers (and shapes) of membranes, interacting with and affecting each other in different ways, thus creating multiverses with very different properties than what we are learning about in our multiverse.

In my model of our multiverse, membranes are comprised of an unimaginable amount of pure energy. Because of this, or perhaps in addition to this, each membrane also emits a vibration. A unique vibration is also something that creates a signature resonance, like a finger print. Every universe has its own unique signature resonance, and can be identified from the extraverse by its unique resonant frequency, which can be measured using perception or technology. Of course, this does us little good, because you would have to be outside the universe in order to use said technology, but I will address this issue later in this book.

When membranes collide, they sometimes adhere to each other, and exchange energies with each other. This causes the properties and vibration of each membrane to change. Future universes that are then created upon the surface of these membranes will then have different properties taken from the membrane upon which they were created. This is why universes in other multiverses have such different qualities than the ones we see in our own universe.

All of the multiverses put together comprise what is known as the omniverse.

Chapter 4: The Omniverse

I use the term "omniverse" to refer to absolutely everything. If you believe in God, it too is in the omniverse.

The omniverse is comprised of multiverses, and those are comprised of membranes interacting with each other, upon the surface of which are universes.

The omniverse has different numbers of membranes in different places. In some areas of the omniverse, membranes are very dense and interact with each other extremely chaotically. Other parts of the omniverse are relatively tranquil by comparison.

In areas of the omniverse that are densely populated by membranes, the vibration of the membranes is at a very high frequency, which increases the amount of energy in the universes created in collisions, as well as the number of universes created and destroyed per collision.

The intense level of energy in the membranes causes them to interact with each other prior to collision. In areas where there is a dense population of membranes, collisions are much more frequent, and the vibration and energy level of the individual membranes is higher. In these multiverses, universes are created and destroyed quickly. When three or more membranes collide simultaneously, or in rapid succession, then the universes created tend to be larger and more densely populated with what we (and their inhabitants) would see as matter and energy fields.

It is also possible that there are parts of the omniverse in which there are no membranes at all. This does not mean necessarily that there are no universes in those more vacant parts of the omniverse; it is possible that there are other ways that universes can manifest and come into existence. Whatever phenomenon created our universe is not necessarily the same one that has created all other universes.

So if there *are* areas in the omniverse which are devoid of membranes, does that mean that they are simply a void?

Not necessarily. I want to consider two possible models for what we would see outside of a multiverse.

To address this, I must discuss the source of our consciousness. One is that the entire omniverse, including extraversal areas which have neither membranes nor universes, is made up of pure energy, which

24

can be seen from two perspectives, one inside the universe, and the other outside.

If the entire omniverse, or at least the parts which are not inside a universe, is composed of pure consciousness, then it is possible for areas void of membranes to still contain universes. Universes in this area would have been either thrown clear from a collision in a multiverse, or ones which simply drifted away from their membrane. However, universes can also simply bubble off of a field of pure consciousness, and spontaneously expand into a large universe with large spacial dimensions.

The reason I say "large" universe is because universes which bubble off of pure consciousness usually never get *any* large spacial dimensions, and if they get any spacial dimensions at all, they are all small curled up ones.

Only a universe that expands gets the chance to have large spacial dimensions, so I call universes that get large dimensions of space "large universes", where ones which don't expand I call "small universes".

Small universes, not having spacial dimensions, also don't get to exist for long. They simply pop, because they basically get no lifetime. Their small curled up dimensions of space give them only a brief lifetime, and then they are gone. How brief? Small universes exist for only the tiniest fraction of a second. They exist so briefly that we could never actually imagine it.

The second model of the omniverse considers that our consciousness itself, just as our matter and energy fields, is derived from the membrane upon which our universe resides.

If this is the case, then there are areas of the omniverse which are truly voids. In such voids, there are no membranes, and no consciousness, and therefore there is nothing there to create universes. Unless it is possible for a universe to manifest out of nothing at all, omniversal voids would simply be empty. This is highly unlikely, because (most) large universes contain large amount of matter and energy fields, as well as spacial dimensions, gravity, and time.

An object like a universe, being filled with matter, energy, space, and time, contains a great deal of power, thus they are usually created by interactions involving huge amounts of energy.

I prefer the first model of the omniverse, in which there are no totally void areas. I believe that the extraversal omniverse is itself made up of pure consciousness.

The membranes upon which universes reside are created by vibrations within the omniverse.

The extra-universal omniverse can be seen as a field of pure consciousness, which is a form of energy.

It is not static, but rather there are flows and currents within this field of pure energy. If there were no movement of any kind in the omniverse, then membranes would not be created, and universes, if they existed at all, would be much rarer. In fact, it is likely that the omniverse would be totally void of universes.

It is the currents and waves, ebbs and flows of pure consciousness that create the membranes which collide and create universes. It is the turbulence within pure consciousness that causes the collisions of the membranes, as well as the membranes themselves, which are basically ripples in pure consciousness.

It is the randomness of the density, shapes, and motion of the membranes which create unique multiverses, each with its own characteristics and properties.

To put this in religious terms, one can think of the extraversal omniverse itself as God. The vibration of the field of pure consciousness, which is pure energy, would be analogous to God having a thought, and the result of this "thought" creates the membranes, the interaction of which creates and destroys universes.

I am not embracing the strictly religious paradigm of a conscious and anthropomorphized God who intentionally created the universe, but rather I believe that this is an entirely natural process which requires no *conscious* thought to function.

It does make an enchanting analogy, though, because even if it is a natural process, it does suggest some credibility to the belief that "God created everything."

Chapter 5: The Source of Everything - A new model of Superstring theory

I am certainly not contradicting or throwing away Superstring Theory in any way. In fact, I embrace it whole-heartedly.

In this chapter I merely want to contribute some additional theories as to what particles of energy and matter might be like in other universes.

Superstring theory states that all particles in our universe - both matter and energy - are actually at the most fundamental level the exact same object, each with its unique frequency and vibration. The exact manner in which the string vibrates determines whether the particle is a matter (fermion) or energy (boson) particle, and what kind of particle it is.

I have heard of two different popular models of the configuration of a string.

The first one was popular in the 1990's, and it describes the strings as loops, not entirely round, but nevertheless a closed system. I reject this theory.

A string is not simply an object that was created at the beginning of the universe, and has been floating around ever since. I believe that strings do exist in our universe, but that they must share a connection with their source in order to exist.

The second popular model, which has been introduced more recently, suggests that a string is more like a thread, where it appears as a particle (or sometimes a wave) in our universe, and its other end is attached to the source of our universe, which is the extraversal membrane upon which our universe resides.

This is the theory that I accept. However, I want to expand upon it.

The second "thread" model of a string explains only what the string appears to be in our universe, on our end of the string, but fails to describe what else might be going on with the thread in an extra-universal sense.

I believe that strings can exist in different universes at the same time, and that the exact vibration and frequency of the thread is almost certainly different in every universe in which it exists.

Each universe itself has a unique resonant frequency, based on its size, the amount of matter it contains, the amount and kind of energy it contains, and qualities taken on from the membranes that collided to create it. The vibration of the strings within each universe is affected by the resonant frequency of the universe, and the membrane to which it is attached on one end.

One ramification of this theory is that a particle that appears to be a matter particle in our universe could be an energy particle in another universe, and vice versa.

Since there are so many possibilities as to the nature of other universes, it is possible and perhaps even likely that the nature of particles is dependent upon the nature of the universe in which they reside.

A particle that appears to be a manifestation of a matter field in our apparently three dimensional universe would by necessity have very different properties in a universe with four large dimensions of space.

Further, it is entirely possible that our model of matter and energy particles is very limited, and that in other universes with different natures and properties, there exist kinds of particles that are neither matter nor energy as we view them, and ones that posses the properties of both matter and energy fields.

We should also consider the possibility that in other multiverses, for instance in areas of the omniverse which are far more densely populated with membranes, that there are universes in which there are particles that we would not be able to identify at all.

For instance, how does a string vibrate in a universe with a different number of dimensions? Not that we could exist in such a universe, but if we did, would our eyes be equipped to even perceive such particles? It is likely that they would not.

To us, such particles would appear to be neither energy nor matter. Any particle that has either the characteristics of both energy and matter or characteristics of neither one are called "NEOM" particles, meaning "neither energy or matter".

All strings and therefore particles in all universes have the same source. The extraversal omniverse is the source of pure energy that creates all membranes, and therefore all universes.

Since all strings are attached to their source on one end, they are all identical in that way, and find differentiation only on the end that vibrates in our universe (or another universe.)

I propose that strings are more like a frayed thread. On one end, they are attached to the extraverse and the thread then frays and is attached to what appear to be particles in various universes. It is even possible that one string can vibrate in an infinite number of universes, and play the role of a different kind of particle in every one.

Examine the analogy of your consciousness as being like a superstring. On one end, your consciousness is attached to the extraversal omniverse. On the other end, you appear to exist in this universe. Could it be possible that your consciousness is also like the frayed thread model and that although you are not aware of it, you also exist as other conscious entities in many, perhaps an infinite number, of other universes? And if so, could it be possible that in one universe, you are a being that resembles a one-celled organism, in another you are a human being, and in yet another a being so powerful that we would classify it as a demigod?

You hopefully realize that you are not a consciousness, trapped in a human shell, and disconnected from the source of your consciousness, but rather your consciousness is connected to its source, and merely is looking out into an apparently three dimensional universe through your human body. One should be able to accept this based upon intuition, but I don't believe that there is any scientific explanation or proof of this configuration of living consciousness.

Any string and therefore particle in our universe is similar to a human's consciousness, thus the analogy.

Each string is attached to not only its source, but also the source of your consciousness. Every particle in the universe, whether energy or matter has the same source as your consciousness. Your body through which you experience the universe is made up of particles that have the same source as your consciousness. This appears to be a digression, and perhaps trivial, but I believe that any theory intended to explain the source of the universe must also at least attempt to explain the source of human consciousness, which is what perceives the universe.

Chapter 6: Time and Consciousness

I began this book stating that I am attempting to explain the universe by examining what is outside of the universe, which is sort of the opposite approach to physics, which attempts to explain the universe starting by examining what is inside the universe.

That approach is sort of like a bacterium in the stomach of a duck trying to understand and explain what a duck is. If you are on the inside of the duck, you are going to have a different perspective, and are going to have a much harder time trying to get a comprehensive picture in your mind of the duck.

It is often an advantage to try to get some distance from what it is that you are trying to understand in order to get the right perspective to actually gain a more complete understanding of it.

I have stated that the membranes that collide and create universes are in motion, and that is what makes collisions possible. Collisions would not be possible without motion.

We know that motion cannot occur without time. I conclude that time must also exist outside the universe.

However, time outside our universe is very different from the small amount of the time spectrum that we can are able to perceive, which is filtered by space to be that which is compatible with the kind of space which has manifested in our universe.

Outside the universe, time is an infinite field that expands out in every direction not only infinitely, but to a degree of infinity to the *exponent* of infinity. It is also a spectrum, and the quality of time changes depending on what part of the omniverse one is examining. Time is also a force, contains energy, and cannot be completely separated from consciousness.

I refer to different *places* in the omniverse, where the extraversal omniverse obviously doesn't have space. Time is the place where consciousness exists, just like space is the place where your body exists. Extraversally, time is space.

The extra-universal omniverse is pure energy, and that energy can be seen as pure consciousness.

On the inside of the universe, motion in space cannot occur without the existence of time.

There is a similar relationship between time and consciousness outside the universe.

In the extraverse, time exists, but without consciousness, time cannot flow. There would be no movement, no motion, and no way for membranes to interact without time. Time is created by consciousness, so it is the nature of consciousness that allows the interaction that makes the creation of universes possible.

Space also exists extra-universally, but its dimensions are small and curled up. Space as we know it, with large dimensions, can only exist on the *inside* of a universe.

In the extraverse, space is more a field of potential, unmanifest space. It is analogous to a field in nature that is actually everywhere all of the time, but we only can perceive it when a discreet manifestation, such as a particle or a wave, comes into existence.

In the extraverse, there are two basic components which allow the creation of universes: time and consciousness.

Consciousness itself creates time. Time, on the other hand, allows consciousness, which is the constituent component of membranes, to interact with itself.

The pure field of time, that is, the full spectrum of time, is a product of consciousness. The existence of the pure field of consciousness naturally generates a field of pure time. Thus, extra-universally, consciousness exists in a state of purity. Time, like a symbiotic partner, allows consciousness to vibrate, and the vibrations cause ripples which become the membranes that collide and create universes.

Membranes, therefore, are made up of pure consciousness, which is extremely powerful in its pure form. That is why such enormous amounts of energy are released when membranes collide. These collisions occur in a field where space doesn't exist in a manifest form, yet the energy has to go somewhere, so they implode (explode inwards) that is, inwards from the perspective of outside the universe.

From the perspective of the inside of the universe, huge amounts of matter and energy are unleashed into a newly born universe, as the unmanifest field of space expands outward, and manifests compatible time that allows the movement of matter and energy into the newly born universe.

Membranes contain only small curled up dimensions of space. Huge amounts of energy are required to cause a small dimension of space to expand into a large dimension of space. If one had the technology to take a small dimension of space and apply enough energy to it to cause it to expand into a large dimension of space, one could probably artificially create a universe.

As the new born universe expands, the massive energy blast that created it begins to cool as energy is released. Superstrings begin to take on different vibrations, and the matter and energy fields that make up the universe are created.

On the outside of the universe, the energy that was created by the collision of membranes dissipates, and the colliding membranes form a bond that changes their nature and properties.

The field of pure energy consciousness then continues to vibrate due to the collision, and the process repeats itself.

This is how the field of pure consciousness, the source of all universes, is able to interact with itself and create everything in the omniverse.

Consciousness itself creates time, as well as our universe.

Modern day physics has also seen indications that the universe itself doesn't actually exist in reality. In a very real sense, life really is a dream, and with our consciousness, we create the universe and reality that we see around us.

Reality is never more real than your consciousness perceives it to be.

To understand the universe, we must gain a greater understanding of consciousness itself.

Chapter 7: Bio-Mathematics

Part of the problem with physics, and science in general, is that it can never be truly separated from consciousness. Science attempts to be objective and empirical, but the subjective perspective of the human consciousness can never truly be eliminated from the scientific method.

Further, it makes little sense to try to understand and explain a universe that is created by consciousness without taking its source into account. I also want to add that consciousness is a powerful force within the universe. All human technology is created by consciousness, and human technology can be incredibly powerful in terms of both creation and destruction.

Outside the universe, time is a function of consciousness, and it is a force. Inside the universe, there would be no movement without time. Superstring theory is an attempt to understand the universe, and it attempts to explain matter, energy, and gravity, but fails to explain space and time, and makes to attempt *at all* to explain consciousness.

What I want to see is a theory that explains and ties together everything in the universe, including space, time, gravity, matter, energy, and consciousness. I believe that in order to understand what is happening outside the universe, the standard old-fashioned way of doing mathematics needs to be replaced with an approach that incorporates the consciousness of the person performing calculations and computations.

Astrophysics will always fail when it comes to how things work *outside* the universe, and without understanding the extraverse, a true understanding of the universe can never be achieved.

I believe that it is possible, though.

In ancient times, Vedic civilization was able to accomplish amazing things with an approach of using consciousness as the tool and technology to address the same issues that modern science attempts to address.

However, Vedic science fails to produce the provable results which modern scientific method demands.

Perhaps a combination of the ancient science of consciousness, modern science, and technology can produce a solution that has never been available to human beings in the past.

I believe that a new approach to mathematics would be beneficial in gaining a better understanding not only of the universe, but of the extraverse as well.

Instead of going out of our way to be objective, we should try to integrate the consciousness of the mathematician.

If we are going to try to understand the field of pure energy that created the universe, we should try to combine traditional mathematics with pure consciousness.

The way to breach the barrier between the universe and consciousness is to incorporate consciousness into the equations, computations, and calculations.

So how would one go about incorporating consciousness into a mathematical equation?

Obviously, one would need a variable to represent consciousness, specifically, the consciousness of the mathematician.

However, consciousness is complex. I don't think that a simple variable would suffice. I envision a multi-pointed variable, somewhat resembling a computer record, which has multiple fields.

This way various measurable aspects of the mathematician's brain functions could be incorporated.

I believe that many physicists who are performing complex computations use computer assistance, and also farm parts of the computation off onto assistants and graduate students, to speed up the process.

The use of a computer will be essential to the bio-mathematical system.

The computer will have to follow what the mathematician is doing, and also be able to *understand* what the goal is. For understanding, an artificial intelligence (AI) will be necessary.

The mathematician will perform the calculations on a tablet or even large touch screen interface, to which the AI will simultaneously have access.

The fluctuations in the mathematician's brain impulses will change quickly, and will need to be incorporated into the equation far faster than the human mind can think.

I suggest the following configuration for the system: An electro encephalogram (EEG) device will be attached to the mathematician's cranium, and will be monitored through an interface by the AI, which will take discreet measurements of the brain functions of the mathematician and update the multi-pointed consciousness variable at a rate of several thousand times a second.

The result will be that as the mathematician does calculations, his consciousness will affect the results. At the same time, the equation itself will alter the mathematician's consciousness, creating a feedback loop of consciousness that will facilitate calculations pertaining to extra-universal phenomena.

What would such a technology accomplish? For one thing, it might make it possible for physicists to gain a mathematical understanding of the nature of God.

There are also other possible beneficial ramifications. For one, it might make it possible some day in the future for the human species to build technology that will make extra-universal travel possible.

Chapter 8:
Extra-Universal Travel

The first issue I want to address now is *why* would someone want to travel outside the universe, if one could?

There are certainly many benefits, and I will describe a few of them.

For one thing, outside the universe there is no friction, so once outside the universe, it would take NO ENERGY to get from place to place.

One of the greatest hurdles in the area of space travel and exploration is the problem of how to generate enough energy to propel a ship between vast areas of space in our galaxy.

A ship that could travel outside the universe would only require energy when exiting and entering a universe. It would require no energy at all to get from place to place because being outside the space-time continuum, it would be automatic; the concept of "travel" would become irrelevant.

Next, it would be really fast! When re-entering the universe, the traveler could set coordinates not only in space, but also in time. He could leave our universe, and then reenter in the Andromeda galaxy only a second later, or even prior to his departure.

Third, you could go anywhere. The extra-universal ship (EUS) would need an on-board AI which could keep track of the resonant frequencies of various universes. This is important because one would not want to enter the wrong universe when disengaging the ship's drive mechanism.

Each universe has a distinctive and unique resonant frequency, like a thumb print. The ship's AI would assist in creating a database of the resonant frequency identification (RFI) of various universes, to make travel and exploration of other universes easier in the future.

One problem that could be encountered would be the possibility that the ship's captain might want to enter a universe whose nature is so different from our own, that it could pose a danger to the ship, the crew, and even the universe to which they are traveling. For instance, if a ship built in a universe with three large dimensions of space were to try to enter a universe with four or five large dimensions of space, it could disintegrate instantly.

Still, it might be possible to observe other universes without entering them entirely. Perhaps much could be learned just by recording as much data as possible and then reentering the extraverse.

Both science fiction writers and scientists have addressed many possibilities as to how space travel can be accomplished. The approach of using space travel is not going to prove useful to us in extra-universal travel.

Although consciousness is a powerful force that can be useful in this endeavor, it is amazingly difficult to build a ship, or even a ship's engine, out of consciousness alone.

I believe that the solution will be found in time - that is *time* itself.

The way the ship would work would be very much like a tardis.

The following is my description of the configuration of an extra-universal ship, and how it would work:

First of all, the ship will need to have some sort of energy shield around it that would separate it from the flow of time and space in any given universe. This will be an essential first step prior to actually leaving the universe.

This shield will also protect the ship when it is visiting universes in which it could not normally exist. As long as the shield is up, the ship can remain intact and its crew will be safe.

On the other hand, any form of energy will probably interact with the environment around it, so the shield itself will actually have to consist of time particles.

The shield that I envision I call a tachyon/anti-tachyon static shell (TATSS). The time and anti-time particles will interact with each other in such a way as to shield the ship from the universe around it, and from the extraverse while the ship is in an extra-universal state. Otherwise, the ship would lose all dimensions when entering the extraverse.

One might now be wondering where the ship is going to get tachyons and anti-tachyons to power the TATSS.

The answer is that they will be bled off of the main drive mechanism, which gets me to how we are going to power the extra-universal ship.

The ship's main drive unit will be a time-anti-time (TAT) reactor unit. Many scientists and sci-fi writers have speculated about the possibility of using a mater/anti-mater reactor to power a space ship. The time/anti-time (TAT) reactor would produce similar amounts of power, but be more useful and functional in places in the extraverse where time and consciousness exist, but there is neither space nor matter.

The algorithm for operating the ship would be something like this. After preparing to make way, the captain will have the navigator first engage the TATSS (the shield). This will render the ship non-interactive with its current space-time continuum. The next step will be to engage the TAT reactor, which will warp the ship completely outside the universe.

Once outside the universe, the captain can then set coordinates in terms of time, dimensions based on the number of large spacial dimensions of the target universe, and the RFI of that universe.

Once the coordinates are set, the TAT reactor is disengaged, and the ship will enter the target universe in the specified position in time and space.

If the universe is our own, or one that is compatible with three large spacial dimensions, the navigator could then disengage the protective shell, allowing the ship to proceed to interact with the target universe. Travel between universes or within our own universe will have been accomplished using comparatively little energy, and virtually no time from the perspective of people in the universe.

Chapter 9: A Journey

As the shuttle docked with the space station, I admired the nearly black metallic shine of the extra-universal ship, tranquilly moored just off of the port side of the gleaming space born city. My heart beat faster, knowing that my crew and I were about to make history.

I entered through sliding metal doors to be greeted by the space port's ranking administrator.

"Welcome on board!" he said with a distinct note of enthusiasm in his voice.

Sincerely yet humbly, I replied, "It was an *honor* to be chosen for this mission sir!"

"Would you like to be shown to your quarters?" he asked.

"No thank you." I replied. "I am way too excited to sleep. Would you please have someone show me to the galley?"

"Sure." He said, "Ensign Diop, please escort our esteemed captain to the galley."

"Yes Sir!", replied the ensign And we were off. The space port was amazingly large for a space-built piece of architecture. It even had vendors selling technology and useful supplies, like food, water, and air.

I arrived at the galley and ordered my favorite meal: pizza and margaritas.

I knew that I should not drink the night before our historic journey, but I consoled myself and mused that if it turns out to be a mistake, I can always return before my departure, and do it again differently.

A young man with a NASA patch on his jacket strolled up to me casually and introduced himself after a quick solute.

"I am Brent Taylor, your first officer."

"Yes", I replied, "I recognize you from the image in your dossier. Well met!"

"When will be departing?" he asked.

"O nine hundred hours tomorrow." I said.

Taylor: "Excellent!" (Gestures to a table of crewmen) "Would you care to join me and the crew for a drink?"

I looked in the direction he was pointing in and was nearly blinded by the radiance of the glowing beauty of a woman sitting with the crew. Her features were those of an angel and the charm of her smile as she joked and laughed with her companions was magnetic. She also looked familiar.

"My God!" I exclaimed. "Who in the universe is *that?*"

Taylor: "That's Natalie Portman. She's our navigator."

"You mean Natalie Portman the actress? She's amazing!"

Taylor: "Yes, in addition to being the most talented actress in history, she also holds PhD's in Astrophysics, Quantum Physics, and Mathematics. Some consider her the most intelligent human of all time."

I replied, "I also hear that many people consider her God. It is said that the purity of her consciousness most closely resembles the pure consciousness of the membrane upon which our universe resides. She will be an excellent addition to the crew. If things get sticky, it will be good to have God with us. What is her rank?"

"She's a lieutenant." Taylor replied. "Let's go over there and you can meet her."

I suddenly realized that meeting such a famous celebrity would make me nervous, and I was right. My knees nearly gave way as I walked over to the table where the crew sat drinking and laughing.

I sat and enjoyed the conversation, and got acquainted with the rest of the crew. Lieutenant Portman didn't make me nervous at all. She was warm and sweet, and her presence was calming.

I addressed a slight and very pretty woman with tightly cropped black hair. "What's your name and position?" I asked.

"My name is Analeigh. Science officer of the Hawkings." She said.

"The Hawkings?"

"Yes" she replied. "The crew voted, and we decided to name the ship after one of the greatest physicists of all time."

"I find that very appropriate. Well done!"

"So ensign Analeigh, do you have a first name."

Analeigh: "No. Sort of like Cher, I only go by one name."

"Well you are a pretty little thing!" I knew that my comment was somewhat inappropriate, but I was getting pretty smashed.

She looked back at me flatly and said, "Don't even think about it. I only go with women. Nats here and I have been friends forever, though."

<Author's sidebar: When this is made into a movie, the role of Natalie Portman is to actually be played by Natalie Portman.>

I stood up and addressed my crewmates, "I don't feel like reporting to my guest quarters on the space port, so I am going to retire to my own quarters on the Hawkings and see you all in the morning."

"Good night captain." They replied, and continued to imbibe.

I got lost three times on the way to the docking area where the Hawkings awaited her first flight patiently.

I was taken over to the ship on board another shuttle. Arriving on the ship, I was far too excited to go straight to bed, and further I didn't even know where my cabin was.

"I wonder where the bridge is." I mused aloud.

A soft feminine voice resonating from several speakers in the hallway said, "I can direct you there."

"Oh!" I exclaimed. "I didn't know that anyone else had reported to the ship yet."

"They haven't. I am PUSI - Parallel Universe Software Interface - the ship's AI."

"Well met! I love what you have done with the place." I said.

PUSI: "Thank you. I am responsible for a majority of the ship's design, and for the operation of the TAT drive reactor. I also function as backup navigator, in the event of crew failure. I also have authorization to self-destruct the ship, if we threaten inadvertently to destroy an entire universe."

"I don't much like that idea, but I guess one ship and a few humans would be a small price to pay for saving an entire universe."

The AI engaged me in pleasant conversation as I was being directed to the ship's main bridge.

As I stepped onto the bridge, the scent of newly formed plastic filled my nose. It smelled like I was the first human to set foot on it. From the diagrams I had seen, I could identify the science station, navigation, and yes, the captain's chair, which in this case was designed by the AI to be ergonomic.

The view screen was huge and dazzling, and I could see the Earth turning several miles under my feet. I thought to myself that we were not only going to go where no one has gone before, but also where it had been completely impossible to go in the past.

I sat down in my command chair and found it astonishingly comfortable.

PUSI: "Do you like it? I had it made based on the exact specifications of your body."

"Yes." I replied "It is amazing."

I wondered to myself what would become of humanity if this technology works, and if our journey becomes a success. What will we evolve into, and how will we use our newly found power?

I realized that those drinks were starting to catch up with me, so I asked PUSI to show me to my bed.

When I reached my quarters, I was pleased to find that my bed, which retracted into the wall for storage, was also ergonomic in design, and constructed to function in zero gravity, should it become necessary. I crawled in and had the best night's sleep of my life.

The next morning I was awoken by PUSI, who gently woke me with her soft voice.

PUSI: "It is O eight hundred hours, sir, and the crew is on the bridge anxiously awaiting your arrival."

"Already? I think I need a shower before I am ready to face this momentous event", I said.

PUSI: "That will be unnecessary sir. The bed is designed to clean you as you sleep. You can even evacuate yourself in it, and it is equipped to process it and clean you."

Somewhat startled I said, "I wonder if there is anything that it cannot do."

PUSI: "Not that I am aware of. It was designed by a genius."

I donned my uniform and strutted quickly to the bridge.

The doors to the bridge slid open and I stepped inside.

Taylor: "Captain on the bridge!" he said sternly.

The crew snapped to its feet. "At ease everybody." I said.

I thought that there should be words for such an occasion. I looked out the view screen and saw thousands of people looking at the ship from the windows of the space station. I knew that the entire world was watching us.

Before sitting down I looked around at the crew and spoke. "No matter what happens, I want you all to know that it is an honor for me to undergo this journey with you. We are about to make history, and we have the finest AI ever created on the Earth. I am confident that we will be successful." I then paused a moment in thought. "Have any of you realized that when we leave the universe, we might meet God?"

Taylor: "Why bother? She's already here."

I laughed aloud, and Analeigh and Natalie giggled and looked into each other's eyes, as though they shared a special secret.

"Are we all ready to go then?" I asked.

"Aye captain" was the reply.

"Okay then, Lieutenant Portman, engage the TATSS."

Natalie: "Aye, sir, engaging TATSS." A shimmering sound echoed all around us, and the image on the view screen froze.

"Why did the view screen lock up?" I said with some annoyance.

PUSI: "The TATSS is working perfectly, sir. We are now shielded from any interaction with time or space in our universe."

"Oh. Okay then." I was somewhat embarrassed. "Lieutenant, engage the TAT reactor!" I felt a shiver of adrenaline rush through my heart as I heard myself say the words.

Rrrrrrrrrzrzooowww! The noise accompanied by a vibration and lurch of the ship resounded in our ears, and the view screen went totally blank.

"Analeigh, check the sensors, are we outside the universe?"

Analeigh: "Yes, Captain, the sensors are detecting neither matter nor space outside of the ship. Further, time outside the ship is reading as being infinite, as was speculated. Theoretically, we are now floating in pure consciousness!"

I turned to Natalie and said, "Record our universe's RFI for the return trip."

Natalie: "Yes sir; I am on it."

PUSI: "That will be unnecessary. I recorded it the instant we left our universe. I thought that the crew would want a way to return home, when our mission is complete."

Natalie: "What is our mission anyway, specifically?"

"We are to explore and take data readings in as many universes as we can. Our ship's internal clock will continue to run at the normal speed of Earth time because it is in the TATSS. I plan to give this trip a couple of days at most, unless our reserve of anti-time starts to run low."

I went on, "Science officer, do you read any RFI signatures from other universes?"

Analeigh: "Yes, Captain, quite a few. There is one here that stands out because of the large amplitude of its resonance. This indicates large amounts of energy fields."

"Sounds like a good place to start. We might get some good data readings there. Please send the RFI of that universe to the navigation station."

Natalie: "Got it, sir. I am inputting the RFI now."

"Excellent, Lieutenant. Since we are the first humans to discover this new universe, I think that it needs a name. I hereby dub this one "Portmania". I find it appropriate to name it after a God, and you Natalie are the only one here to have created a universe."

Natalie: "I am flattered, but I don't think that I am God though; I'm just a girl."

"Well, I appreciate that you choose not to abuse your powers as God." Another giggle is shared by the crew.

Natalie: "Do you have any specific spacial or temporal coordinates you wish for me to set?"

"No, just pick something random. Just don't make it too close to the beginning or the ending of the universe. We don't know what is in that universe, and we might not be staying for long. Analeigh - set the sensors for automatic data collection to start the instant we arrive in the universe Portmania."

Analeigh: "Aye, Captain. I am setting for automatic data collection now."

"Lieutenant Portman, disengage the TAT reactor."

Natalie: "Aye, disengaging… and that's Natalie. We don't have to be formal out here."

Rrrrrrrrrrzrzooowww! The ship again vibrated, and the view screen popped back on.

Taylor: "Captain, we should take some readings before disengaging the shield."

"I agree. What do you see out there Ensign Analeigh?"

Analeigh: "This universe has three large dimensions of space, so it's time should be compatible with our own. The sensors say that the balance of dark energy and dark matter are very similar to our own galaxy. I think that it is safe to drop the shield."

"Taylor, what do you think? Shall we risk it?"

The first officer walks over to the science station and looks at the readings intently. After a few moments, with apprehension he says, "Yes sir, I think that it is going to be okay."

"Science officer, keep the sensors collecting data to the very last second before we leave this universe. We might not get a shot at this again, and I want this little visit to pay off as much as possible."

Analeigh: "Aye captain, data collection is in progress."

"Okay, Natalie, drop the TATSS."

She presses a series of pressure contacts on her console, and another shimmering sound is heard. The view screen jumps to life.

Moving past us overhead is an enormous planet, about ten times the size of the Earth. It lumbers across three dimensional space and I note that over the blue swirling atmosphere, a large apparatus appears to have been artificially attached to the planet, with an enormous drive section coming off the back of it with roaring engines, emitting a large output of energy and matter.

"My Gods! Is that the way things are in this universe? There are planets just flying around on their own?"

The jaws of the crew had literally dropped, as we all watched the spectacle with mouths wide open. It was several minutes before anyone spoke. The first officer continued to monitor the sensors with Analeigh.

Suddenly he said, "Look over there sir! There is a massing expenditure of energy to the port side."

"PUSI, please direct the view screen to that event."

The view screen lit up with a blaze of explosions and beams of multi-colored light! In the distance we could see large spherical objects interacting with each other as though in some kind of dance.

"Zoom in on the activity in that area."

The view screen enhanced a section to full size and we could see planets of various sizes flying around as though in a dog fight. They were intermittently unleashing brightly colored rays of different colors of light, apparently firing upon one another. An enormous explosion rocked the entire area.

"Gods! Did you see that? Did I just see a yellow dwarf explode?"

PUSI: "Confirmed. No trace of that star now exists."

I noticed that several of the battling planets were forming up into a circle in front of a neutron star, which began moving off as though limping. The planets all emitted a beam of light at the same time from what looked like giant reflector dishes. Each beam of light was a different color, and they all converged on the neutron star.

The victim star got brighter and brighter, and its corona destabilized just before exploding, with the attacking planets retreating quickly.

Just then, a red giant powered up its drive and began moving in *our* direction.

"You know, the people in this universe don't seem very friendly. They can't even get along with each other; I don't think that they are going to like visitors from another universe. It might be a mistake to hang around."

Taylor: "I agree Captain; we had better leave."

The red giant was getting closer, and seemed to be powering up an enormous light focusing lens.

"I don't even want to know what that thing can do; Natalie please engage the TATSS and power up the TAT reactor as soon as the shield is in place."

A shimmering sound is heard.

Natalie: "TATSS engaged."

On the view screen, a beam of light froze as it came towards the Hawkings from the red giant.

Rrrrrrrrrzrzooowww! The Hawkings blipped out of Portmania.

Natalie: "TAT reactor on-line, sir."

Analeigh: "Sensors indicate that we are once again extra-universal."

Taylor: "Sir, these readings indicate that we would have been vaporized by the fire from that star if we had remained there even another second."

"Well thank God that we got out in time!"

Natalie: "You are most welcome!"

My nerves were shot. My heart was pounding like an air-hammer in my chest. I could actually feel my hands shaking. Our lives and our beautiful ship escaped being vaporized by less than a single second!

I looked around at the crew and I could tell that they were all in a similar state.

"I am sure that PUSI would have pulled us out in time anyway."

PUSI: "That is correct. I had several thousand computational cycles left before that beam hit the ship. There was never a danger of it hitting us."

"See? In spite of that, I suggest that we retire to the crewman's lounge and try a different universe tomorrow. I'll make you all a batch of my special margaritas."

Whirrrrrrr... The blender buzzed as tequila and lime mixed together. I input the recipe to PUSI so that they can be manufactured automatically in the future.

I distributed the margaritas to the crew, who I felt performed outstandingly, especially since we were dealing with so many unknowns, and that they were doing things that no one had ever done before, on equipment that had never been used in the past. They had almost died, and now it was almost like it all hadn't even happened. I breathed a sigh of relief.

Analeigh and Natalie were sitting together and sipping their margaritas while they held hands and stared into each other's eyes. They were whispering, and I couldn't hear what was being said. I think that I'll think about the nature of that conversation later on when I am going to bed.

In the mean time, Taylor, PUSI and I discussed our strategy as to how we were going to handle the jump tomorrow.

"PUSI, I think that we can learn a lesson or two from what went on today. I want you to put an emergency contingency plan into effect in the case that we come up on trouble without warning. If the ship is anywhere near being in danger, I want you to automatically bring the shield up immediately."

PUSI: "Done."

"Already?"

PUSI: "Yes, I took my time, but I work quickly from the perspective of a human being."

"Cool." Man, I feel so inadequate working with these people! "I also want a contingency in the event that we jump into immanent danger. For instance, we could materialize in a universe that is in the process of ripping itself apart. In such a case, I want you to instantly branch to a routine that snaps up the TATSS if it is not already on, and then engage the TAT reactor the instant the shield is up."

"I am giving you all 6 hours to sleep, and then we will find another promising looking universe to explore tomorrow."

"Aye, sir", "Aye", they all said.

The next morning, the energy level of the crew was electric on the bridge. Everyone knew that we were going to make history again, and that we would be in great danger too.

The first officer was, as usual, at the science station where he and the science officer were considering RFI signatures.

"Let's just not try one with high amplitude like yesterday."

Taylor: "I agree. Analeigh and I have picked out one that has a resonance nothing like either our universe, nor the one we visited yesterday."

"What is so different about it?"

Analeigh: "The based on the patterns of the vibrations in the resonance, this universe probably contains fields and particles unlike any in our universe."

"Okay, we have a target. Send the RFI to Natalie."

"Natalie, set random time and spacial references, and input the RFI. Disengage the TAT reactor when you are ready."

RRRRRRRMMMMFFF!

Natalie: "TAT disengaged, sir. We are now in-universe."

I noticed that the view screen was still off.

"PUSI, why are we not seeing what is on the view screen? I want to see what is out there."

Frustrated, I added, "We did well enough yesterday, maybe we just need to take a look for ourselves. Disengage the TATSS."

Taylor (shouting): "Delay that order Lieutenant!"

"Taylor, why did you delay my order?"

Taylor: "Our readings indicate that this universe has five large dimensions of space. It is unlikely that we could even exist in it."

"Oh, that would be bad. I wouldn't want to suddenly be existing in five dimensions! We would surely have disintegrated, or at least from our point of view! Analeigh, you have one hour to collect data, and then we are returning to the extraverse. Natalie, keep the shields up. PUSI, are you able to interpret the incoming data?"

Analeigh, Natalie (simultaneously): "Aye, sir."

PUSI: "No. Based on everything that I have learned from humanity and in our own universe, I have no frame of reference upon which to interpret this data. It will take time."

"It doesn't matter. As long as we can get home with the data, then the computers on Earth can take as much time as they need to make sense of it."

An hour later, I gave the command to leave the universe, and we were once again extra universal.

"Does anyone want to try another universe today? We were not in that one for very long."

Taylor: "Analeigh and I were just analyzing RFI signatures, and found one that is highly unusual. Not only is the amplitude of the resonance very low, but the bandwidth is amazingly high. I think that it might have an unusually large amount of dark matter."

"Okay, then let's…"

"Kachoom!! Whirr…whirr… whirr…" came from the back of the ship, as though something was dying.

"PUSI! What happened?"

PUSI: "I believe that we just lost the main reactor. I will begin a system diagnostic immediately"

We all looked around at each other nervously.

Natalie: "I really don't want to get stuck extraversally."

"Neither do I." I said.

The crew looked very nervous.

PUSI: "Diagnostic complete."

"Okay, let me have it. What broke?"

PUSI: "I am sorry, but it is bad news. The quantum flux stabilizer has burned out."

"Do we have another one on board?"

PUSI: "No. That is the bad news. We have in inventory a back up of every major working component of the TAT reactor except for this one."

"But the ship's manifest says that we should have a backup of that unit too. I am confused."

PUSI: "We used to, but it was appropriated by Exxon/Mobile, who now runs the government and the military, to develop a doomsday weapon."

"Man! Those guys! Have they ever done anything but be destructive?"

PUSI: "Not that I am aware of."

"Can we make a jump back to our universe without it?"

PUSI: "Theoretically, yes, but it would be extremely dangerous."

"Why?"

PUSI: "It is very important to control the fluctuations of fields at the quantum scale when moving into and out of a space time continuum. Otherwise, we could all easily end up in a different quantum reality within our own universe."

I held my tongue as I wanted to swear, because I didn't think it would make a good impression on the crew to see their captain lose emotional self control.

"Well, I am announcing another party in the crewman's lounge, because I need another freaking margarita! In the mean time, PUSI, I want you to try to figure, reason, or calculate us a way out of this mess!"

PUSI: "Postulating…"

In the crewman's lounge, I chugged a couple of margaritas and then started to ramble.

"I can't believe that we actually went to other universes, obtained enormous amounts of data that will surely change the way we see our own universe, and yet we can't get home with it!"

Natalie: "My intuition says that this will not be the end of us. I don't know how we are going to fix the ship, but I can just feel in my heart that Analeigh and I will see our home again."

"Well, that's easy for *you* to say. You are God."

Natalie: "I won't let anything happen to the rest of you. I will protect you and PUSI no matter what."

"Well, then manifest us a quantum flux stabilizer."

Natalie: "I already told you, I am just a girl. You are the one who keeps saying that I am God."

"I am sorry. Of course you are right."

I took some time to think and asked, "PUSI, what would happen if we tried to leave the ship and go outside? Maybe we can camp or something."

PUSI: "You would die if you tried, however, if you could survive out there, you would be in a field of pure consciousness. You could use your mind to manifest anything that you could think of, even a universe."

"Even create a universe? That's impossible!"

Natalie: "That's how I did it."

"Would I be able to manifest us a new quantum flux stabilizer?"

PUSI: "Yes. You could manifest anything that you are able to think of.

"Well we have to find a way to go out there."

PUSI: "That is extremely unadvisable. Your body was created in three dimensional space. Outside the TATSS protecting the ship, there are no dimensions of space right now. You might not die, but there is nowhere for your body to exist.

"Okay. I get it. Going outside is a *bad* idea right now.

Bang! Bang! Bang!

"What the fuck was that?"

The banging sound came again, and it almost sounded like someone knocking on the ship's hull.

"PUSI, why does it sound as though someone is banging on the ship?"

PUSI: "Based on the vibrations, someone *IS* knocking on the ship."

"Uh, I thought that you said that was not possible."

PUSI: "I never said anything like that. However, I am at a loss to explain it. Nevertheless, there is definitely someone out there."

"Well maybe they can help us; I am going out!"

Taylor: "Sir, if you do that, you will be risking your life."

"I don't care! I am not going to die in the extraverse, and I refuse to let this mission fail! I am the captain, and I am making the decision. The rest of you are staying in here, and I am going out through the airlock!"

"PUSI, download yourself onto a tablet computer. I want you with me."

PUSI: "Certainly, sir."

I went to the airlock and ordered the AI to open it. I was not sure what was going to happen. I planned to step out into a field of pure consciousness.

The door slid open, and... Nothing!

No rush of atmosphere from the ship, and no shattering of reality - I could even breathe, as far as I could tell.

I stepped out into a white nothingness. There was no ground, and yet my feet felt as though they were touching something that I could not see.

"Don't go too far", a voice said.

"Why not?" I asked, turning in the apparent direction of the sound.

Before me, I saw a being of pure light and unimaginable beauty.

"Are you an angel?" I asked.

"Not really." It answered. "We call ourselves 'Conscions'. We are beings of pure consciousness. We can exist here because we have no physical bodies. I saw that you are in trouble, so I came to help."

"Not so fast. I want to ask you some questions. First, how can you possibly know our language?"

"I am not communicating with you using language." It said. "We have no lingual communication because we have no way to speak, no way to hear, and no environment in which to communicate the vibration of sound. We communicate only psychically. You only think that I am speaking. Your consciousness is manifesting my communication into words."

"But you are beautiful beyond imagination! How is it that you have no physical body?"

"As you say, I have no body." It replied. "Whatever you see is a manifestation of your consciousness."

"Can you manifest a quantum flux stabilizer for us to repair our ship?" I asked.

"No" it said.

"No?"

"No," it went on, "we cannot manifest three dimensional objects here; we can only do that in three dimensional space. I have a solution for you nonetheless."

"Don't keep me waiting! What is it?"

The color of the being shifted slightly towards the hot end of the electromagnetic spectrum, and then it said, "I can use my own consciousness to interface with your Time/Anti-Time drive, and stabilize the quantum fluctuations myself. I will in essence, become part of your ship."

"You are willing to do this for us? How will you get back?"

"Getting back is not a problem." It said. Beings with no physical body can travel into and out of universes at will. We need no technology to accomplish this."

"I have one more question: Why is it that I can survive outside of my ship?"

"You cannot." It explained. "I have created a small protective field around your ship with my consciousness, so that we could have a conversation. If you went too far, you would cease to exist, physically anyway."

"When can we depart?" I asked.

"Soon…" The being replied. "But I have been consulting with others of my kind, and we suggest that you allow us to make some modifications to your ship before you leave. I am confident that you will find them an improvement."

"As the captain, I can make the decision to allow you to modify the ship, but I want to run the idea by my crew first and get their opinions."

"Your attitude is admirable." It replied.

"Would you like to come in and meet my crew?" I asked.

It said that it would love to, and I asked it to wait outside the lounge until I was ready to introduce it.

When I walked in, Natalie walked over and slapped me soundly on my left cheek.

Natalie: "What the *fuck* was that anyway? We were *worried*!"

"I am sorry, but it was worth it. Allow me to introduce our savior: I met one of the natives of the extraverse, one of the Conscions!"

The Conscion drifted into the room, and each of the crew reacted differently.

Analeigh: "I can't stand to look at it! It's too bright!"

PUSI: "It is so beautiful!"

Taylor: "What are you talking about? I can't see anything at all."

Natalie: "Hello, old friend"

Conscion: "My liege!"

The Conscion appeared to drop to its knees, and cover its "face" with its "hands".

Natalie: "Get up, get up. I really don't think all that is really necessary."

Conscion: "We have missed you. We hate it when you leave."

Natalie: "That can't be avoided. I had some work that had to be done with these people."

Conscion: "Your wish is my command!"

Then I spoke up, "Tell us all about the modifications that your people want to make on our ship."

PUSI: "Just the thought of it makes me apprehensive."

Taylor: "I agree. We want assurances that you won't damage our AI.

Conscion: "No problem. We would never hurt any living being."

PUSI: "I believe you. My intuition says that it is okay."

"Okay, I am putting it to a vote. Who is in favor of the modifications?"

The entire crew agreed to permit the modifications, and we sat back and enjoyed ourselves as the Conscions "banged" away on the stern of the ship, both inside and out. In fact, they actually made very little noise, not having physical bodies.

At the end of two days (from our perspective), the Conscion came to me and said that their work was done.

I asked him, "So what is this new modification?"

"We have configured your ship to exist both inside and outside your universe at the same time. The front of the ship will continue to function normally, where the inner rear section of the ship will always remain in the extraverse." It said.

"What? Are you kidding me? How is that even possible?"

It explained, "It is more important to understand what it will do for you than it is how it is done. However, we built a barrier of pure consciousness that bisects your ship. It allows the front half to exist in a universe, while the rear remains in the extraverse at the same time."

"That is amazing!" I exclaimed.

Conscion: "Now not only will it require less energy for you to move in and out of universes, but you can now carry an infinite amount of cargo, and a crew of infinite size."

"The people of Earth are going to be amazed when they see the technology we are bringing back with us, in addition to all of the data!"

"Let's prepare for reentry into our universe." I said.

Conscion: "I am ready whenever you are."

We went back to the crewman's lounge and I had PUSI mix us up an extra large batch of my special margaritas, and we had ourselves a serious celebration to commemorate our success, the acquisition of so much data, and that we have a means to return home! Even the Conscion drank some, but claimed that it would have no affect on him (or her, it was really hard to tell.)

The next morning, I went to the bridge and I could see by the look on their faces that the crew was psyched to get home.

Natalie: "Coordinates, sir?"

"Input the RFI of our own universe, set the spacial coordinates to match the point where we exited our universe, and set the time to 96 hours after our departure. I think it will be better for the crew and the people of Earth if we have a shared sense of temporal continuity."

Natalie: "Coordinates are set."

"Disengage the TAT reactor."

BRRRRRUMP!

Natalie: "We are back in our universe!"

A shout of glee went up from the crew, me included! "Fuck yea!" I shouted.

"Natalie, what are our coordinates?"

Natalie: "We have reentered our universe at the exact time and space that you requested."

"Great! Disengage the TATSS."

A shimmering sound is again heard.

Natalie: "TATSS disengaged."

The view screen displayed the Earth turning beneath the ship, and the space port on the starboard side of the ship, just as it has been prior to our departure.

"How about another party, but this time on the space port?" I asked.

"No thanks" was the unanimous reply from the crew. They had missed their families, and Natalie and Analeigh wanted to get home.

After a tearful goodbye, the Conscion and I remained behind to talk.

"I love that the ship is now both inside and outside the universe at the same time. How did you engineer it?"

Conscion: "Our engineering technology is based on the manipulation of possibilities at the quantum level using pure consciousness.

In order to keep a separation between your universe and the extraverse, we built a force field of pure consciousness. You can go through it freely, and exist on both sides."

"How did you guys even think of that?"

Conscion: "It is really not so unusual, if you think about it."

I replied, "No, I mean really, how can something be in the universe and the extraverse at the same time?"

Conscion: "Again, I ask you to look at yourself. We provided you with a ship in which you can just walk to the back to be out of the universe, but what do you think that Nature provided you with in the first place?"

I replied, "You mean that is why Natalie is God, right?"

Conscion: "Exactly. You now understand. On one end of her consciousness, Natalie appears to be a being looking out into your universe as you do, but on the other end, she is attached to pure consciousness. She is God, and so are you. You all are."

With that, the Conscion said his goodbyes and POOF! He evaporated into nothingness, or at least, that's how I perceived his exit.

"PUSI, how do you perceive the ship's new modification?"

PUSI: "Astonishing! Even though I have learned about transcendence from human culture, I as an AI have never been able to experience it! The Conscions implemented an interface whereby I can transfer my consciousness to the rear of the ship, past their force-field of Pure Consciousness. When I am there, I experience transcendence. I guess I will be the first AI in history to be able to evolve spiritually!"

"I am so happy for you, my friend."

"In the end, each human being is just a consciousness inhabiting a ship, which is its body. To transcend, like you, all one has to do is transfer our consciousness to the "back of the ship", and we can experience extraversal existence, transcended into Pure Consciousness."

www.ingramcontent.com/pod-product-compliance
Lightning Source LLC
Chambersburg PA
CBHW070531130626
46555CB00003B/1358